COMMERCIAL PILOT

Practical Test Standards

for

GLIDER

FLIGHT STANDARDS SERVICE
Washington, DC 20591

FOREWORD

The Commercial Pilot – Glider Practical Test Standards (PTS) book has been published by the Federal Aviation Administration (FAA) to establish the standards for commercial pilot certification practical tests for the glider category. FAA inspectors and designated pilot examiners shall conduct practical tests in compliance with these standards. Flight instructors and applicants should find these standards helpful during training and when preparing for practical tests.

/s/ 9/22/2006

Joseph K. Tintera, Manager
Regulatory Support Division, AFS-600
Flight Standards Service

The following major enhancements were made to FAA-S-8081-2

- The reference section has been updated to current FAA publications in use.
- An abbreviation section has been added.
- The "plan of action" is more thoroughly explained to include the requirement to use scenarios.
- Practical test prerequisite requirements have been updated to current rules.
- The list of Special Emphasis Items has been updated.
- Unsatisfactory Performance has been clarified.
- The use of Letters of Discontinuance has been described.
- The requirement to evaluate the applicant's Aeronautical Decision Making and Risk Management skills has been addressed.
- The need for testing in Single-Pilot Resource Management has been addressed and clarified.
- Reference to the metric system has been eliminated.
- The body of the PTS has been updated to include airworthiness requirements, self-launch systems, visual signals, runway incursion avoidance, launches and landings, and slow flight.

NOTE

Material in FAA-S-8081-23A will be effective November 1, 2006. All previous editions of the Commercial Pilot -Glider Practical Test Standards will be obsolete as of this date.

RECORD OF CHANGES

IV. AREA OF OPERATION: LAUNCHES AND LANDINGS

G. TASK: Aerotow – Abnormal Occurrences

Q. TASK: Landings – Normal and Crosswind Landing

- To clarify the intent, extent, and condition of the evaluation.

Contents

INTRODUCTION 10

General Information 10

Practical Test Standard Concept 11

Practical Test Book Description 11

Abbreviations 12

Special Emphasis Areas 14

Commercial Pilot – Glider Practical Test Prerequisites 15

Aircraft and Equipment Required for the Practical Test 16

Flight Instructor Responsibility 16

Examiner Responsibility 16

Satisfactory Performance 17

Unsatisfactory Performance 17

Letter of Discontinuance 18

Aeronautical Decision Making and Risk Management 19

Single-Pilot Resource Management 19

Applicant's Use of Checklists 19

Use of Distractions during Practical Tests 19

Positive Exchange of Flight Controls 20

CHECKLISTS:

Applicant's Practical Test Checklist 21

Examiner's Practical Test Checklist 23

ADDITIONAL RATING TASK TABLE 26

AREA OF OPERATION:

I. PREFLIGHT PREPARATION 27

A. Certif cates and Documents 27

B. Airworthiness Requirements 27

C. Weather Information 28

D. Operation of Systems 28

E. Performance and Limitations 29

F. Aeromedical Factors 30

AREA OF OPERATION:

II. PREFLIGHT PROCEDURES 31

A. Assembly 31

B. Ground Handling 31

C. Preflight Inspection 32

D. Cockpit Management 32

E. Visual Signals 32

III. AIRPORT AND GLIDERPORT OPERATIONS 33

A. Radio Communications and ATC Light Signals 33

B. Traffic Patterns 33

C. Airport, Runway, and Taxiway Signs, Markings, and Lighting 34

IV. LAUNCHES AND LANDINGS 35

AEROTOW

A. Before Takeoff Check 35

B. Normal and Crosswind Takeoff 35

C. Maintaining Tow Positions 36

D. Slack Line 36

E. Boxing the Wake 36

F. Tow Release 37

G. Abnormal Occurrences 37

GROUND TOW (AUTO OR WINCH)

H. Before Takeoff Check 38

I. Normal and Crosswind Takeoff 38

J. Abnormal Occurrences 39

SELF-LAUNCH

K. Engine Starting 39

L. Taxiing 39

M. Before Takeoff Check 40

N. Normal and Crosswind Takeoff and Climb 40

O. Engine Shutdown In Flight 41

P. Abnormal Occurrences 41

LANDINGS

Q. Normal and Crosswind Landing 42

R. Slips to Landing 43

S. Downwind Landing 43

V. PERFORMANCE SPEEDS 44

A. Minimum Sink Airspeed 44

B. Speed-To-Fly 44

VI. SOARING TECHNIQUES 45

A. Thermal Soaring 45

B. Ridge and Slope Soaring 45

C. Wave Soaring 46

VII. PERFORMANCE MANEUVERS 47

A. Straight Glides 47

B. Turns to Headings 47

C. Steep Turns 48

VIII. NAVIGATION 49

A. Flight Preparation and Planning 49

B. National Airspace System 50

IX. SLOW FLIGHT AND STALLS 51

A. Maneuvering at Minimum Control Airspeed
AIRSPEED 51
B. Stall Recognition and Recovery 52
X. EMERGENCY OPERATIONS 53
A. Simulated Off-Airport Landing... 53
B. Emergency Equipment and Survival Gear 53
XI. POSTFLIGHT PROCEDURES 54
A. After-Landing and Securing 54

INTRODUCTION

General Information

The Flight Standards Service of the Federal Aviation Administration (FAA) has developed this practical test book as the standard that shall be used by FAA inspectors and designated pilot examiners[1] when conducting commercial pilot - glider practical tests. Flight instructors are expected to use this book when preparing applicants for practical tests. Applicants should be familiar with this book and refer to these standards during their training.

The FAA gratefully acknowledges the valuable assistance provided by many industry participants who contributed their time and talent with the revision of these practical test standards.

---

1 The word "examiner" denotes either the FAA Inspector, FAA designated pilot examiner, or other authorized person who conducts the practical test.

Practical Test Standard Concept

Title 14 of the Code of Federal Regulations (14 CFR) part 61 specifies the areas in which knowledge and skill must be demonstrated by the applicant before the issuance of a commercial pilot certificate or rating. The CFRs provide the flexibility to permit the FAA to publish practical test standards containing the AREAS OF OPERATION and specific TASKs in which pilot competency must be demonstrated. The FAA will revise this PTS whenever it is determined that changes are needed in the interest of safety. ***Adherence to the provisions of the regulations and the practical test standards is mandatory for evaluation of commercial pilot applicants.***

Practical Test Book Description

This test book contains the commercial pilot - glider practical test standards. This includes the AREAS OF OPERATION and TASKs for the issuance of an initial commercial pilot certificate and for the addition of category ratings to that certificate.

AREAS OF OPERATION are phases of the practical test arranged in a logical sequence within each standard. They begin with Preflight Preparation and end with Postflight Procedures. The examiner may conduct the practical test in any sequence that results in a complete and efficient test; ***however, the ground portion of the practical test shall be accomplished before the flight portion.***

TASKs are titles of knowledge areas, flight procedures, or maneuvers appropriate to an AREA OF OPERATION.

NOTE is used to emphasize special considerations required in the AREA OF OPERATION or TASK.

REFERENCE identifies the publication(s) that describe(s) the TASK. Descriptions of TASKs are not included in the standards because this information can be found in the current issue of the listed references. Publications other than those listed may be used for references if their content conveys substantially the same meaning as the referenced publications.

These practical test standards are based on the following references

14 CFR part 21	Certification Procedures for Products and Parts
14 CFR part 43	Maintenance, Preventive Maintenance, Rebuilding, and Alteration
14 CFR part 61	Certification: Pilots, Flight Instructors and Ground Instructors
14 CFR part 91	General Operating and Flight Rules

FAA-H-8083-1	Aircraft Weight and Balance Handbook
FAA-H-8083-13	Glider Flying Handbook
FAA-H-8083-25	Pilot's Handbook of Aeronautical Knowledge
AC 00-6	Aviation Weather
AC 00-45	Aviation Weather Services
AC 21.17-2	Type Certification—Fixed-Wing Gliders (Sailplanes), including Powered Gliders
AC 60-22	Aeronautical Decision Making
AC 60-28	English Language Skill Standards Required by 14 CFR parts 61, 63, and 65
AC 61-65	Certification: Pilots and Flight and Ground Instructors
AC 61-67	Stall and Spin Awareness Training
AC 61-84	Role of Preflight Preparation
AC 61-134	General Aviation Controlled Flight into Terrain Awareness
AC 90-48	Pilots' Role in Collision Avoidance
AC 90-66	Recommended Standards Traffic Patterns for Aeronautical Operations at Airports without Operating Control Towers
AC 120-51	Crew Resource Management Training
AIM	Aeronautical Information Manual
AFD	Airport Facility Directory
NOTAMs	Notices to Airmen
Others	Pertinent Pilot's Operating Handbooks (POH)/Glider Flight Manuals (GFM)

The Objective lists the important elements that must be satisfactorily performed to demonstrate competency in a TASK. The Objective includes:

1. specifically what the applicant should be able to do;
2. the conditions under which the TASK is to be performed; and
3. the acceptable standards of performance.

Abbreviations

14 CFR	Title 14 of the Code of Federal Regulations
AC	Advisory Circular

ADM	Aeronautical Decision Making
ATC	Air Traffic Control
CFIT	Controlled Flight into Terrain
CRM	Cockpit Resource Management
FAA	Federal Aviation Administration
FSDO	Flight Standards District Office
GFM	Glider Flight Manual
GPO	Government Printing Office
LAHSO	Land and Hold Short Operations
NAVAID	Navigation Aid
NOTAM	Notice to Airmen
NWS	National Weather Service
POH	Pilot's Operating Handbook
PTS	Practical Test Standard

Use of the Practical Test Standards

The practical test standards for commercial pilot - glider are designed to evaluate competency in both knowledge and skill.

The FAA requires that all practical tests be conducted in accordance with the appropriate practical test standards and the policies set forth in the INTRODUCTION. Applicants shall be evaluated in **ALL** TASKs included in the AREAS OF OPERATION of the appropriate practical test standard (unless noted otherwise).

An applicant who holds a commercial pilot certificate and is seeking an additional glider category rating, will be evaluated on at least the AREAS OF OPERATION and TASKs listed in the Additional Rating Task Table located on page 1-vii of this practical test standard. At the discretion of the examiner, an evaluation of the applicant's competence in the remaining AREAS OF OPERATION and TASKs may be conducted.

In preparation for each practical test, the examiner shall develop a written "plan of action" for each practical test. The "plan of action" is a tool, for the sole use of the examiner, to be used in evaluating the applicant. The plan of action need not be grammatically correct or in any formal format. The plan of action must contain all of the required AREAS OF OPERATION and TASKs and any optional TASKs selected by the examiner.

The "plan of action" shall incorporate one or more scenarios that will be used during the practical test. The examiner should try to include as many of the TASKs into the scenario portion of the test as possible, but maintain the flexibility to change due to unexpected situations as they arise and still result in an efficient and valid test. If the elements in one TASK have already been evaluated in another TASK, they need not be repeated. ***Any TASK selected for evaluation during a practical test shall be evaluated in its entirety.***

The examiner is not required to follow the precise order in which the AREAS OF OPERATION and TASKs appear in this book. The examiner may change the sequence or combine TASKs with similar Objectives to have an orderly and efficient flow of the practical test. For example, Boxing the Wake may be combined with Maintaining Tow Positions. The examiner's "plan of action" shall include the order and combination of TASKs to be demonstrated by the applicant in a manner that will result in an efficient and valid test.

All TASKs in these practical test standards are required for the issuance of a commercial pilot – glider certificate, regardless of launch method, except for AREA OF OPERATION IV. Per the first Note in AREA OF OPERATION IV, the examiner must select the kind of launch, based on the applicant's qualifications. In AREA OF OPERATION IV the kind of launch is shown in each TASK heading.

When using the practical test standards, the examiner must evaluate the applicant's knowledge and skill in sufficient depth to determine that the standards of performance listed for all applicable TASKs are met. However, when a particular element is not appropriate to the glider, its equipment, or operational capability, that element may be omitted. Examples of these element exceptions would be gliders not equipped with variometers or total energy compensators, avionics, or electrical systems.

The examiner is expected to use good judgment in the performance of simulated emergency procedures. The use of the safest means for simulation is expected. Consideration must be given to local conditions, both meteorological and topographical, at the time of the test, as well as the applicant's workload, and the condition of the aircraft used. If the procedure being evaluated would jeopardize safety, it is expected that the applicant shall simulate that portion of the maneuver.

Special Emphasis Areas

Examiners shall place special emphasis upon areas of aircraft operations considered critical to flight safety. Among these are:

1. positive aircraft control;
2. positive exchange of the flight controls procedure;
3. stall/spin awareness;
4. collision avoidance;

5. wake turbulence avoidance;
6. LAHSO;
7. runway incursion avoidance;
8. CFIT;
9. ADM and risk management;
10. checklist usage;
11. temporary flight restrictions (TFR);
12. special use airspace (SUA);
13. aviation security; and
14. other areas deemed appropriate to any phase of the practical test.

Although these areas may not be specifically addressed under each TASK, they are essential to flight safety and will be evaluated during the practical test. In all instances, the applicant's actions will relate to the complete situation.

Commercial Pilot – Glider Practical Test Prerequisites

An applicant for a Commercial Pilot - Glider practical test is required by 14 CFR part 61 to:

1. be at least 18 years of age;
2. be able to read, speak, write, and understand the English language; (If there is doubt, use AC 60-28, English Language Skill Standards).
3. hold at least a private pilot certificate issued under 14 CFR 61;
4. have passed the commercial pilot - glider knowledge test since the beginning of the 24th month before the month in which the practical test is completed;
5. have satisfactorily accomplished the required training and obtained the prescribed aeronautical experience;
6. present the Airman Knowledge Test Report, if required, at the time of application for the practical test;
7. have a written endorsement from an authorized flight instructor certifying that the applicant has been given flight training in preparation for the practical test within 60 days preceding the date of application and is prepared to pass the practical test; and

8. have a written endorsement from an authorized flight instructor certifying that the applicant has satisfactory knowledge of the subject area(s) in which a deficiency was indicated by the Airman Knowledge Test Report.

Aircraft and Equipment Required for the Practical Test

The commercial pilot - glider applicant is required by 14 CFR part 61 to provide an airworthy, certificated aircraft for use during the practical test. This section further requires that the aircraft must:

1. be of U.S., foreign or military registry of the same category, class, and type for the certificate and/or rating for which the applicant is applying;
2. have fully functioning dual controls, except as provided for in 14 CFR sections 61.45(c) and (e); and
3. be capable of performing all AREAS OF OPERATION appropriate to the rating sought and have no operating limitations which prohibit its use in any of the AREAS OF OPERATION required for the practical test.

Flight Instructor Responsibility

An appropriately rated flight instructor is responsible for training the commercial pilot applicant to acceptable standards in all subject matter areas, procedures, and maneuvers included in the TASKs within the appropriate practical test standard.

Because of the impact of their teaching activities in developing safe, proficient pilots, flight instructors should exhibit a high level of knowledge, skill, and the ability to impart that knowledge and skill to students. Additionally, the flight instructor must certify that the applicant is able to perform safely as a commercial glider pilot and is competent to pass the required practical test.

Throughout the applicant's training, the flight instructor is responsible for emphasizing the performance of effective visual scanning, collision avoidance, and runway incursion avoidance procedures. These areas are covered, in part, in AC 90-48, Pilot's Role in Collision Avoidance; FAA-H8083-13, Glider Flying Handbook; FAA-H-8083-25, Pilot's Handbook of Aeronautical Knowledge; and the Aeronautical Information Manual.

Examiner Responsibility

The examiner conducting the practical test is responsible for determining that the applicant meets the acceptable standards of knowledge and skill of each TASK within the appropriate practical test standard. Since there is no formal division between the "oral" and "skill" portions of the practical test, this becomes an ongoing process throughout the test. To avoid unnecessary distractions, oral questioning should be used judiciously at all times, especially during the flight portion of the practical test.

Examiners shall test to the greatest extent practicable to the applicant's correlative abilities rather than mere rote enumeration of facts throughout the practical test.

If the examiner determines that a TASK is incomplete, or the outcome uncertain, the examiner may require the applicant to repeat that TASK, or portions of that TASK. This provision has been made in the interest of fairness and does not mean that instruction, practice, or the repeating of an unsatisfactory TASK is permitted during the certification process.

Throughout the flight portion of the practical test, the examiner shall evaluate the applicant's use of visual scanning and collision avoidance procedures.

The examiner may not assist the applicant in the management of the aircraft, radio communications, navigational equipment, and navigational charts.

The examiner shall remain alert for other traffic at all times. The examiner shall use proper ATC terminology when simulating ATC clearances.

Satisfactory Performance

Satisfactory performance to meet the requirements for certification is based on the applicant's ability to safely:

1. perform the TASKs specified in the AREAS OF OPERATION for the certificate or rating sought within the approved standards;
2. demonstrate mastery of the aircraft with the successful outcome of each TASK performed never seriously in doubt;
3. demonstrate satisfactory proficiency and competency within the approved standards; and
4. demonstrate sound judgment and ADM.

Unsatisfactory Performance

The tolerances represent the performance expected in good flying conditions. If, in the judgment of the examiner, the applicant does not meet the standards of performance of any TASK performed, the associated AREA OF OPERATION is failed and therefore, the practical test is failed.

NOTE: The tolerances stated in this standard are intended to be used as a measurement of the applicant's ability. They provide guidance for examiners to use in judging the applicant's qualifications. The regulations governing the tolerances for operation under Visual Flight Rules are established in 14 CFR part 91.

The examiner or applicant may discontinue the test at any time when the failure of an AREA OF OPERATION makes the applicant ineligible for the certificate or rating sought. ***The test may be continued ONLY with the consent of the applicant.*** If the test is discontinued, the applicant is entitled credit for only those AREAS OF OPERATION and their associated TASKs satisfactorily performed. However, during the retest, and at the discretion of the examiner, any TASK may be reevaluated, including those previously passed.

Typical areas of unsatisfactory performance and grounds for disqualification are:

1. Any action or lack of action by the applicant that requires corrective intervention by the examiner to maintain safe flight.
2. Failure to use proper and effective visual scanning techniques to clear the area before and while performing maneuvers.
3. Consistently exceeding tolerances stated in the Objectives.
4. Failure to take prompt corrective action when tolerances are exceeded.

When a disapproval notice is issued, the examiner shall record the applicant's unsatisfactory performance in terms of the AREA OF OPERATION and specific TASKs not meeting the standard appropriate to the practical test conducted. The AREA(s) OF OPERATION/TASKs not tested and the number of practical test failures shall be recorded. If the applicant fails the practical test because of a special emphasis area, the Notice of Disapproval shall indicate the associated TASK. For example, AREA OF OPERATION II, TASK C, Preflight Inspection, failure to inspect the glider using the appropriate checklist.

Letter of Discontinuance

When a practical test is discontinued for reasons other than unsatisfactory performance (i.e., equipment failure, weather, or illness) FAA Form 8700-1, Airman Certificate and/or Rating Application, and, if applicable, the Airman Knowledge Test Report shall be returned to the applicant. The examiner at that time shall prepare, sign, and issue a Letter of Discontinuance to the applicant. The Letter of Discontinuance should identify the AREAS OF OPERATION and their associated TASKs of the practical test that were successfully completed. The applicant shall be advised that the Letter of Discontinuance shall be presented to the examiner when the practical test is resumed, and made part of the certification file.

Aeronautical Decision Making and Risk Management

The examiner shall evaluate the applicant's ability throughout the practical test to use good aeronautical decision making procedures in order to evaluate risks. The examiner shall accomplish this requirement by developing scenarios that incorporate as many TASKs as possible to evaluate the applicant's risk management in making safe aeronautical decisions. For example, the examiner may develop a scenario that incorporates weather decisions and performance planning.

The applicant's ability to utilize all the assets available in making a risk analysis to determine the safest course of action is essential for satisfactory performance. The scenarios should be realistic and within the capabilities of the aircraft used for the practical test.

Single-Pilot Resource Management

Single-Pilot Resource Management refers to the effective use of ALL available resources: human resources, hardware, and information. It is similar to Crew Resource Management (CRM) procedures that are being emphasized in multi-crewmember operations except that only one crewmember (the pilot) is involved. Human resources "...includes all other groups routinely working with the pilot who are involved in decisions that are required to operate a flight safely. These groups include, but are not limited to: tow operators, ground crew (such as wingwalkers), weather briefers, maintenance personnel, and air traffic controllers." Pilot Resource Management is not a single TASK; it is a set of skill competencies that must be evident in all TASKs in this practical test standard as applied to single-pilot operation.

Applicant's Use of Checklists

Throughout the practical test, the applicant is evaluated on the use of an appropriate checklist. Proper use is dependent on the specific TASK being evaluated. The situation may be such that the use of the checklist, while accomplishing elements of an Objective, would be either unsafe or impracticable, especially in a single-pilot operation. In this case, a review of the checklist after the elements have been accomplished would be appropriate. Division of attention and proper visual scanning should be considered when using a checklist.

Use of Distractions during Practical Tests

Numerous studies indicate that many accidents have occurred when the pilot has been distracted during critical phases of flight. To evaluate the pilot's ability to utilize proper control technique while dividing attention both inside and/or outside the cockpit, the examiner shall cause a realistic distraction during the flight portion of the practical test to evaluate the applicant's ability to divide attention while maintaining safe flight.

Positive Exchange of Flight Controls

During flight, there must always be a clear understanding between pilots of who has control of the aircraft. Prior to flight, a briefing should be conducted that includes the procedure for the exchange of flight controls. A positive three-step process in the exchange of flight controls between pilots is a proven procedure and one that is strongly recommended.

When one pilot wishes to give the other pilot control of the aircraft, he or she will say, "You have the flight controls." The other pilot acknowledges immediately by saying, "I have the flight controls." The first pilot again says "You have the flight controls." When control is returned to the first pilot, follow the same procedure. A visual check is recommended to verify that the exchange has occurred. There should never be any doubt as to who is flying the aircraft.

APPLICANT'S PRACTICAL TEST CHECKLIST

Commercial Pilot—Glider

EXAMINER'S NAME______________________________

LOCATION____________________________________

DATE/TIME___________________________________

ACCEPTABLE AIRCRAFT

- □ Aircraft Documents:
  - Airworthiness Certificate
  - Registration Certificate
  - Operating Limitations
- □ Aircraft Maintenance Records:
  - Record of Airworthiness Inspections
  - Current Status of Applicable Airworthiness Directives
- □ Pilot's Operating Handbook and FAA-Approved Glider Flight Manual

PERSONAL EQUIPMENT

- □ Practical Test Standard
- □ Current Aeronautical Charts
- □ Computer and Plotter
- □ Flight Plan Form
- □ Flight Log Form
- □ Current AIM, Airport Facility Directory, and Appropriate Publications

PERSONAL RECORDS

- □ Identification - Photo/Signature ID
- □ Pilot Certificate
- □ Completed FAA Form 8710-1, Airman Certificate and/or Rating Application with Instructor's Signature (if applicable)

- ☐ Airman Test Report
- ☐ Pilot Logbook with Appropriate Instructor Endorsements
- ☐ FAA Form 8060-5, Notice of Disapproval (if applicable)
- ☐ Approved School Graduation Certificate (if applicable)
- ☐ Examiner's Fee (if applicable)

EXAMINER'S PRACTICAL TEST CHECKLIST

Commercial Pilot—Glider

APPLICANT'S NAME ______________________________

LOCATION ______________________________

DATE/TIME ______________________________

I. PREFLIGHT PREPARATION

- □ **A.** Certificates and Documents
- □ **B.** Weather Information
- □ **C.** Operation of Systems
- □ **D.** Performance and Limitations
- □ **E.** Aeromedical Factors

II. PREFLIGHT PROCEDURES

- □ **A.** Assembly
- □ **B.** Ground Handling
- □ **C.** Preflight Inspection
- □ **D.** Cockpit Management
- □ **E.** Visual Signals

III. AIRPORT AND GLIDERPORT OPERATIONS

- □ **A.** Radio Communications
- □ **B.** Traffic Patterns
- □ **C.** Airport, Runway, and Taxiway Signs, Markings, and Lighting

IV. LAUNCHES AND LANDINGS

AEROTOW

- □ **A.** Before Takeoff Check
- □ **B.** Normal and Crosswind Takeoff

- ☐ **C.** Maintaining Tow Positions
- ☐ **D.** Slack Line
- ☐ **E.** Boxing the Wake
- ☐ **F.** Tow Release
- ☐ **G.** Abnormal Occurrences

GROUND TOW (AUTO OR WINCH)

- ☐ **H.** Before Takeoff Check
- ☐ **I.** Normal and Crosswind Takeoff
- ☐ **J.** Abnormal Occurrences

SELF-LAUNCH

- ☐ **K.** Engine Starting
- ☐ **L.** Taxiing
- ☐ **M.** Before Takeoff Check
- ☐ **N.** Normal and Crosswind Takeoff and Climb
- ☐ **O.** Engine Shutdown in Flight
- ☐ **P.** Abnormal Occurrences

LANDINGS

- ☐ **Q.** Normal and Crosswind Landing
- ☐ **R.** Slips to Landing
- ☐ **S.** Downwind Landing

V. PERFORMANCE AIRSPEEDS

- ☐ **A.** Minimum Sink Airspeed
- ☐ **B.** Speed-To-Fly

VI. SOARING TECHNIQUES

- ☐ **A.** Thermal Soaring

□ **B.** Ridge and Slope Soaring

□ **C.** Wave Soaring

VII.PERFORMANCE MANEUVERS

□ **A.** Straight Glides

□ **B.** Turns to Headings

□ **C.** Steep Turns

VIII. NAVIGATION

□ **A.** Flight Preparation and Planning

□ **B.** National Airspace System

IX. SLOW FLIGHT AND STALLS

□ **A.** Maneuvering at Minimum Control Airspeed

□ **B.** Stall Recognition and Recovery

X. EMERGENCY OPERATIONS

□ **A.** Simulated Off-Airport Landing

□ **B.** Emergency Equipment and Survival Gear

XI. POSTFLIGHT PROCEDURES

□ **A.** After-Landing and Securing

ADDITIONAL RATING TASK TABLE ADDITION OF A GLIDER RATING TO AN EXISTING COMMERCIAL PILOT CERTIFICATE									
Required TASKS are indicated by either the TASK letter(s) that apply(s) or an indication that all or none of the TASKS must be tested.									
AREA OF OPER-ATION	ASEL	ASES	AMEL	AMES	RH	RG	PL	Bal-loon	Air-ship
I	B,C,D, E	B,C,D, E	B,C,D, E	B,C,D, E	B,C,D, E	B,C,D, E	B,C,D, E	B,C,D, E	B,C,D, E
II	A,B,C, E	A,B,C, E	A,B,C, E	A,B,C, E	A,B,C, E	A,B,C, E	A,B,C, E	ALL	A,B,C, E
III	B	B	B	B	B	B	B	ALL	B
IV	ALL*	ALL*	ALL*	ALL*	ALL*	ALL*	ALL*	ALL*	ALL*
V	ALL	ALL	ALL	ALL	ALL	ALL	ALL	ALL	ALL
VI	ALL	ALL	ALL	ALL	ALL	ALL	ALL	ALL	ALL
VII	ALL	ALL	ALL	ALL	ALL	ALL	ALL	ALL	ALL
VIII	NONE	NONE	NONE	NONE	NONE	NONE	NONE	A	NONE
IX	ALL	ALL	ALL	ALL	ALL	ALL	ALL	ALL	ALL
X	ALL	ALL	ALL	ALL	ALL	ALL	ALL	ALL	ALL
XI	ALL	ALL	ALL	ALL	ALL	ALL	ALL	ALL	ALL

*EXAMINER SHALL SELECT KIND OF LAUNCH BASED ON THE APPLICANT'S QUALIFICATIONS.

LEGEND

ASEL	Airplane Single-Engine Land
ASES	Airplane Single-Engine Sea
AMEL	Airplane Multiengine Land
AMES	Airplane Multiengine Sea
RH	Rotorcraft Helicopter
RG	Rotorcraft Gyroplane
PL	Powered Lift

I. AREA OF OPERATION: PREFLIGHT PREPARATION

A. TASK: CERTIFICATES AND DOCUMENTS

NOTE: The examiner shall develop a scenario to evaluate TASKs C and E. Real time weather or current weather should be used, as available.

REFERENCES: 14 CFR parts 21, 43, 61, 91; AC 61-65; FAA-H8083-1, FAA-H-8083-13, FAA-H-8083-25; POH/GFM.

Objective. To determine that the applicant exhibits knowledge of the elements related to certificates and documents by:

1. Explaining—
    a. commercial pilot certificate privileges, limitations, and recent flight experience requirements.
    b. medical fitness.
    c. pilot logbook or flight records.
2. Locating and explaining—
    a. airworthiness and registration certificates.
    b. operating limitations, placards, instrument markings, and POH/GFM.
    c. weight and balance data and equipment list.

B. TASK: AIRWORTHINESS REQUIREMENTS

REFERENCES: 14 CFR parts 21, 43, 91; AC 21.17-2A; FAA-H 8083-13, FAA-H-8083-25.

Objective. To determine that the applicant exhibits knowledge of the elements related to airworthiness requirements by:

a. Explaining—
    a. required instruments and equipment for day/night VFR.
    b. procedures and limitations for determining airworthiness of the glider with inoperative instruments and equipment.
    c. requirements and procedures for obtaining a special flight permit.

b. Locating and explaining—

a. airworthiness directives.
b. compliance records.
c. maintenance/inspection requirements.
d. appropriate record keeping.

C. TASK: WEATHER INFORMATION

REFERENCES: AC 00-6, AC 00-45, AC 61-84; FAA-H-8083-13, FAAH-8083-25.

Objective. To determine that the applicant:

1. Exhibits knowledge of the elements related to weather information from various sources with emphasis on—
   a. use of weather reports, charts, and forecasts.
   b. significant weather prognostics.
2. Exhibits knowledge of the relationship of the following factors to the lifting process—
   a. pressure and temperature lapse rates.
   b. atmospheric instability.
   c. thermal index and thermal production.
   d. cloud formation and identification.
   e. frontal weather.
   f. other lifting sources.
3. Explains hazards associated with flight in the vicinity of thunderstorms.
4. Makes a competent "go/no-go" decision based on available weather information.

D. TASK: OPERATION OF SYSTEMS

REFERENCES: FAA-H-8083-13, FAA-H-8083-25; POH/GFM.

Objective. To determine that the applicant:

1. Exhibits knowledge of the elements related to the operation of instruments and systems on the glider provided for the practical test, by explaining at least five (5) of the following systems—
   a. magnetic compass.

b. yaw string or inclinometer.
c. airspeed indicator and altimeter.
d. variometer and total energy compensators.
e. gyroscopic instruments.
f. electrical, including starting system for self- launch.
g. landing gear and brakes.
h. avionics.
i. high-lift and drag devices.
j. oxygen equipment.
k. powerplant and propeller for self-launch.
l. fuel, oil and hydraulic for self-launch.

2. Correctly interprets information displayed on the instruments.

E. TASK: PERFORMANCE AND LIMITATIONS

REFERENCES: FAA-H-8083-1, FAA-H-8083-13, FAA-H-8083-25; POH/GFM.

Objective. To determine that the applicant:

1. Exhibits knowledge of the elements related to performance and limitations by explaining the use of charts, tables, and data to determine performance and the adverse effects of exceeding limitations.
2. Demonstrates use of the appropriate performance charts, tables, and data.
3. Computes weight and balance. Determines if the computed weight and center of gravity are within the glider's operating limitations and if the weight and center of gravity will remain within limits during all phases of flight.
4. Explains the management of ballast and its effect on performance.
5. Describes the effect of various atmospheric conditions on the glider's performance.
6. Explains the applicable performance speeds and their uses.
7. Describes the relationship between airspeeds and load factors.

F. TASK: AEROMEDICAL FACTORS

REFERENCES: AIM; FAA-H-8083-13, FAA-H-8083-25; AC 60-22.

Objective. To determine that the applicant exhibits knowledge of the elements related to aeromedical factors explaining:

1. The symptoms, causes, effects, and corrective actions of at least four (4) of the following—
   a. hypoxia.
   b. hyperventilation.
   c. middle ear and sinus problems.
   d. spatial disorientation and illusions.
   e. motion sickness.
   f. carbon monoxide poisoning (self-launch).
   g. stress and fatigue.
   h. dehydration and heatstroke.
2. The effects of alcohol, drugs, and over-the-counter medications.
3. The effects of excess nitrogen during scuba dives upon a pilot or passenger in flight.

II. AREA OF OPERATION: PREFLIGHT PROCEDURES

A. TASK: ASSEMBLY

NOTE: If, in the judgment of the examiner, the demonstration of the glider assembly is impractical, competency may be determined by oral testing.

REFERENCES: FAA-H-8083-13; Glider Flight Manual.

Objective. To determine that the applicant:

1. Exhibits knowledge of the elements related to assembly procedures.
2. Selects a suitable assembly area and provides sufficient crewmembers for assembly.
3. Follows an appropriate checklist.
4. Uses proper tools.
5. Handles components properly.
6. Cleans and lubricates parts, as appropriate.
7. Accounts for all tools and parts at the completion of assembly.
8. Performs post-assembly inspection, including a positive control check.

B. TASK: GROUND HANDLING

REFERENCES: FAA-H-8083-13; Glider Flight Manual.

Objective. To determine that the applicant:

1. Exhibits knowledge of the elements related to ground handling procedures.
2. Selects the appropriate ground handling procedures and equipment for existing conditions.
3. Determines the number of crewmembers needed.
4. Handles the glider in a manner that will not result in damage during movement.
5. Secures the glider and controls, as necessary, in proper position.

C. TASK: PREFLIGHT INSPECTION

REFERENCES: FAA-H-8083-13; Glider Flight Manual.

Objective. To determine that the applicant:

1. Exhibits knowledge of the elements related to preflight inspection, including which items must be inspected, for what reasons, and how to detect possible defects.
2. Inspects the glider using the appropriate checklist.
3. Verifies the glider is in condition for safe flight, notes any discrepancies, and determines if maintenance is required.
4. Inspects the launch equipment, including towline, tow hitches, weak links, and release mechanism.

D. TASK: COCKPIT MANAGEMENT

REFERENCES: 14 CFR part 91; AC 120-51; FAA-H-8083- 13; Glider Flight Manual.

Objective. To determine that the applicant:

1. Exhibits knowledge of the elements related to cockpit management procedures.
2. Organizes and arranges material and equipment in a manner making items readily available.
3. Briefs passengers on the use of safety belts, shoulder harnesses, and emergency procedures.
4. Utilizes all appropriate checklists.

E. TASK: VISUAL SIGNALS

REFERENCE: FAA-H-8083-13.

Objective. To determine that the applicant:

1. Exhibits knowledge of the elements related to aerotow or ground tow visual signals, as appropriate.
2. Uses, interprets, and responds to prelaunch, launch, airborne, and emergency signals, as appropriate.
3. For aerotow, exhibits knowledge of the elements related to in-flight aerotow visual signals, both to and from the towplane.

III. AREA OF OPERATION: AIRPORT AND GLIDERPORT OPERATIONS

A. TASK: RADIO COMMUNICATIONS AND ATC LIGHT SIGNALS

NOTE: If radio communications are impractical, competency may be determined by oral testing.

REFERENCES: AFD; AIM; FAA-H-8083-25.

Objective. To determine that the applicant:

1. Exhibits knowledge of the elements related to radio communications, radio failure, and ATC light signals.
2. Selects appropriate frequencies for facilities to be used.
3. Transmits using recommended phraseology.
4. Acknowledges radio communications and complies with instructions.
5. Uses appropriate procedures for simulated radio communications failure.
6. Interprets and complies with ATC light signals.

B. TASK: TRAFFIC PATTERNS

REFERENCES: 14 CFR part 91; AFD; AC 90-66; FAA-H 8083-13, FAA-H-8083-25.

Objective. To determine that the applicant:

1. Exhibits knowledge of the elements related to traffic pattern procedures for gliders.
2. Follows established traffic pattern procedures.
3. Maintains awareness of other traffic in pattern.
4. Maintains proper ground track with crosswind correction, if necessary.
5. Crosses designated points at appropriate altitudes, unless conditions make such action impractical
6. Selects touchdown and stop points.

7. Adjusts glidepath and track promptly to compensate for unexpected lift, sink, or changes in wind velocity.

8. Makes smooth, coordinated turns with a bank angle not to exceed 45° when turning final approach.

9. Adjusts flaps, spoilers, or dive brakes, as appropriate.

10. Recognizes and makes appropriate corrections for the effect of wind.

11. Completes the prescribed checklist, if applicable.

C. TASK: AIRPORT, RUNWAY, AND TAXIWAY SIGNS, MARKINGS, AND LIGHTING

REFERENCES: 14 CFR part 91; NOTAMS; AFD; AIM; FAA-H8083-25.

Objective. To determine that the applicant:

1. Exhibits knowledge of the elements related to airport, runway, and taxiway operations with emphasis on runway incursion avoidance.

2. Properly identifies, interprets, and complies with airport, runway, and taxiway signs, markings, and lighting.

IV. AREA OF OPERATION: LAUNCHES AND LANDINGS

NOTE: Examiner shall select kind of launch based on the applicant's qualifications.

A. TASK: AEROTOW - BEFORE TAKEOFF CHECK

REFERENCES: FAA-H-8083-13; Glider Flight Manual.

Objective. To determine that the applicant:

1. Exhibits knowledge of the elements related to the before takeoff check, including the reasons for checking the items, and how to detect malfunctions.
2. Establishes a course of action with crewmembers, including signals, speeds, wind, and emergency procedures.
3. Ensures that the glider is in safe operating condition.
4. Checks towline hookup and release mechanism, using the appropriate hook for the type of launch conducted.
5. Ensures no conflict with traffic prior to takeoff.
6. Completes the prescribed checklist, if applicable.

B. TASK: AEROTOW - NORMAL AND CROSSWIND TAKEOFF

NOTE: If a crosswind condition does not exist, the applicant's knowledge of crosswind elements shall be evaluated through oral testing.

REFERENCES: FAA-H-8083-13; Glider Flight Manual.

Objective. To determine that the applicant:

1. Exhibits knowledge of the elements related to normal and crosswind takeoff, including configurations and tow positions.
2. Uses proper pre-launch signals for aerotow launch.
3. Lifts off at an appropriate airspeed.
4. Maintains proper position until towplane lifts off.
5. Maintains directional control and proper wind-drift correction throughout the takeoff.

6. Maintains proper alignment with the towplane.

7. Uses proper aerotow visual signals between the glider and towplane, as appropriate.

C. TASK: AEROTOW - MAINTAINING TOW POSITIONS

REFERENCE: FAA-H-8083-13.

Objective. To determine that the applicant:

1. Exhibits knowledge of the elements related to high-tow (slightly above the wake) and low-tow (slightly below the wake) positions during various phases of aerotow.

2. Makes smooth and correct control applications to maintain vertical and lateral positions during high and low tow.

3. Transitions from high- to low-tow position through the wake while maintaining positive control.

4. Maintains proper tow position during turns.

5. Uses aerotow visual signals as appropriate and as directed by the examiner.

D. TASK: AEROTOW - SLACK LINE

REFERENCE: FAA-H-8083-13.

Objective. To determine that the applicant:

1. Exhibits knowledge of the elements related to the causes, hazards, and corrections related to slack line.

2. Recognizes slack line and applies immediate, positive, and smooth corrective action to eliminate slack line in various situations.

E. TASK: AEROTOW - BOXING THE WAKE

REFERENCE: FAA-H-8083-13.

Objective. To determine that the applicant:

1. Exhibits knowledge of the elements related to boxing the wake (maneuvering around the wake).

2. Maneuvers the glider, while on tow, slightly outside the towplane's wake in a rectangular, box-like pattern.

3. Maintains proper control and coordination.

F. TASK: AEROTOW - TOW RELEASE

REFERENCE: FAA-H-8083-13.

Objective. To determine that the applicant:

1. Exhibits knowledge of the elements related to tow release, including related safety factors.
2. Maintains high-tow position with normal towline tension.
3. Clears the area before releasing the towline.
4. Releases the towline and confirms release by observing the towline.
5. Makes level or climbing turn.

G. TASK: AEROTOW - ABNORMAL OCCURRENCES

REFERENCE: FAA-H-8083-13.

Objective. To determine that the applicant:

1. Exhibits knowledge of the elements related to aerotow abnormal occurrences, for various situations, such as—
   a. towplane power loss during takeoff.
   b. towline break.
   c. towplane power failure at altitude.
   d. glider release failure.
   e. glider and towplane release failure **(oral only)**.
   f. canopy opening in flight.
2. Demonstrates simulated aerotow abnormal occurrences as required by the examiner.

H. TASK: GROUND TOW (AUTO OR WINCH) BEFORE TAKEOFF CHECK

REFERENCES: FAA-H-8083-13; Glider Flight Manual.

Objective. To determine that the applicant:

1. Exhibits knowledge of the elements related to the before takeoff check, including the reasons for checking the items, and how to detect malfunctions.
2. Establishes a course of action with crewmembers, including signals, speeds, wind direction, and emergency procedures.
3. Ensures glider is in safe operating condition.
4. Checks towline hookup and release mechanism, using the appropriate hook for the type of launch conducted.
5. Ensures no conflict with traffic prior to takeoff.
6. Completes the prescribed checklist, if applicable.

I. TASK: GROUND TOW (AUTO OR WINCH) - NORMAL AND CROSSWIND TAKEOFF

NOTE: If a crosswind condition does not exist, the applicant's knowledge of crosswind elements shall be evaluated through oral testing.

REFERENCES: FAA-H-8083-13; Glider Flight Manual.

Objective. To determine that the applicant:

1. Exhibits knowledge of the elements related to normal and crosswind takeoff, including related safety factors.
2. Uses proper signals for takeoff.
3. Maintains directional control during launch.
4. Lifts off at the proper airspeed.
5. Establishes proper initial climb pitch attitude.
6. Takes prompt action to correct high speed, low speed, or porpoising.
7. Maintains proper ground track during climb.
8. Releases in proper manner and confirms release.

J. TASK: GROUND TOW (AUTO OR WINCH) - ABNORMAL OCCURRENCES

REFERENCES: FAA-H-8083-13, Glider Flight Manual.

Objective. To determine that the applicant:

1. Exhibits knowledge of the elements related to ground tow abnormal occurrences for various situations, such as—
   a. overrunning the towline.
   b. towline break.
   c. inability to release towline.
   d. over and under speeding.
   e. porpoising.
   f. canopy opening in flight.
2. Demonstrates simulated ground tow abnormal occurrences, as required by the examiner.

K. TASK: SELF-LAUNCH - ENGINE STARTING

REFERENCE: FAA-H-8083-13, Glider Flight Manual.

Objective. To determine that the applicant:

1. Exhibits knowledge of the elements related to engine starting, including various atmospheric conditions, and awareness of other persons and property during start.
2. Accomplishes recommended starting procedures.
3. Completes appropriate checklists.

L. TASK: SELF-LAUNCH - TAXIING

REFERENCE: FAA-H-8083-13, Glider Flight Manual.

Objective. To determine that the applicant:

1. Exhibits knowledge of the elements related to taxiing, including the effect of wind during taxiing and appropriate control positions.
2. Performs a brake check immediately after the glider begins moving.
3. Positions flight controls properly, considering the wind.

4. Controls direction and speed without excessive use of brakes.
5. Avoids other aircraft and hazards.
6. Complies with signals.

M. TASK: SELF-LAUNCH - BEFORE TAKEOFF CHECK

REFERENCE: FAA-H-8083-13, Glider Flight Manual.

Objective. To determine that the applicant:

1. Exhibits knowledge of the elements related to the before takeoff check, including the reasons for checking each item and to detect malfunctions.
2. Positions the glider properly considering other aircraft, wind, and surface conditions.
3. Ensures engine temperatures and pressures are suitable for run-up and takeoff.
4. Accomplishes before takeoff checks and ensures the glider is in safe operating condition.
5. Reviews airspeeds, takeoff distance, and emergency procedures.
6. Completes appropriate checklists.

N. TASK: SELF-LAUNCH - NORMAL AND CROSSWIND TAKEOFF AND CLIMB

NOTE: If a crosswind condition does not exist, the applicant's knowledge of crosswind elements shall be evaluated through oral testing.

REFERENCE: FAA-H-8083-13, Glider Flight Manual.

Objective. To determine that the applicant:

1. Exhibits knowledge of the elements related to normal and crosswind takeoff and climb.
2. Positions flight controls for existing wind conditions.
3. Clears the area, taxies into takeoff position, and aligns the glider for departure.
4. Advances throttle smoothly to takeoff power.
5. Rotates at recommended airspeed, and accelerates to appropriate climb speed, ±5 knots.

6. Maintains takeoff power to a safe maneuvering altitude, and then sets climb power.

7. Completes appropriate checklists.

O. TASK: SELF-LAUNCH - ENGINE SHUTDOWN IN FLIGHT

REFERENCE: FAA-H-8083-13, Glider Flight Manual.

Objective. To determine that the applicant:

1. Exhibits knowledge of the elements related to engine shutdown procedures in flight.
2. Sets power for proper engine cooling.
3. Establishes appropriate airspeed.
4. Sets electrical equipment.
5. Shuts down engine.
6. Feathers or positions propeller and stows, as applicable.
7. Selects proper static source, if applicable.
8. Completes appropriate checklists.

P. TASK: SELF-LAUNCH - ABNORMAL OCCURRENCES

REFERENCES: FAA-H-8083-13; Glider Flight Manual.

Objective. To determine that the applicant:

1. Exhibits knowledge of the elements related to self-launch abnormal occurrences, for various situations, such as—
    1. partial, complete power failure, and failure to gain restart.
    2. fire or smoke.
    3. electrical system malfunction.
    4. low fuel pressure.
    5. low oil pressure.
    6. engine overheat.

7. canopy opening in flight.

8. engine restart in flight.

2. Demonstrates simulated self-launch abnormal occurrences, as required by the examiner.

Q. TASK: LANDINGS - NORMAL AND CROSSWIND LANDING

NOTE: If a crosswind condition does not exist, the applicant's knowledge of crosswind elements shall be evaluated through oral testing.

REFERENCES: FAA-H-8083-13; Glider Flight Manual.

Objective. To determine that the applicant:

1. Exhibits knowledge of the elements related to a normal and crosswind approach and landing procedures.
2. Adjusts flaps, spoilers, or dive brakes, as appropriate.
3. Maintains recommended approach airspeed, ±5 knots.
4. Maintains crosswind correction and directional control throughout the approach and landing.
5. Makes smooth, timely, and positive control application during the roundout and touchdown.
6. Touches down smoothly within the designated landing area, with no appreciable drift, and with the longitudinal axis aligned with the desired landing path, stopping short of and within 100 feet of a designated point.
7. Maintains control during the after-landing roll.
8. Completes appropriate checklists.

NOTE: The applicant shall touchdown and roll to a point designated by the examiner stopping within 100' without rolling past the designated point. The point should be far enough away from the touchdown point that is should not require more than light-medium braking to come to a stop within the required distance.

R. TASK: LANDINGS - SLIPS TO LANDING

NOTE: The examiner will select one type of slip from the knowledge area for demonstration.

REFERENCES: FAA-H-8083-13; Glider Flight Manual.

Objective. To determine that the applicant:

1. Exhibits knowledge of the elements related to forward, side, and turning slips to landing, with and without the use of drag devices.
2. Recognizes the situation where a slip should be used to land in a desired area.
3. Establishes a slip without the use of drag devices.
4. Maintains the desired ground track.
5. Maintains proper approach attitude.
6. Makes smooth, proper, and positive control applications during recovery from the slip.
7. Touches down smoothly within the designated landing area.

S. TASK: LANDINGS - DOWNWIND LANDING

REFERENCES: FAA-H-8083-13; Glider Flight Manual.

Objective. To determine that the applicant:

1. Exhibits knowledge of the elements related to downwind landings, including safety related factors.
2. Adjusts flaps, spoilers, or dive brakes, as appropriate.
3. Maintains recommended approach airspeed, ±5 knots.
4. Uses proper downwind landing procedures.
5. Maintains proper directional control during touchdown and rollout.
6. Applies brake smoothly to bring glider to a stop.

V. AREA OF OPERATION: PERFORMANCE SPEEDS

A. TASK: MINIMUM SINK AIRSPEED

REFERENCES: FAA-H-8083-13; Glider Flight Manual.

Objective. To determine that the applicant:

1. Exhibits knowledge of the elements related to aerodynamic factors and use of minimum sink airspeed.
2. Determines the minimum sink airspeed for a given situation and maintains the selected speed, ±5 knots.

B. TASK: SPEED-TO-FLY

REFERENCES: FAA-H-8083-13; Glider Flight Manual.

Objective. To determine that the applicant:

1. Exhibits knowledge of the elements related to speed-to-fly, and its uses.
2. Determines the speed-to-fly for a given situation and maintains the airspeed, ±5 knots.

VI. AREA OF OPERATION: SOARING TECHNIQUES

NOTE: Due to varying geographical locations and atmospheric conditions, the applicant may be asked to demonstrate at least one of the following soaring TASKS most appropriate for the particular location and existing conditions. If conditions do not permit a demonstration of soaring skills, applicants will be expected to demonstrate knowledge of the various types of soaring through oral testing.

A. TASK: THERMAL SOARING

REFERENCES: FAA-H-8083-13; AC 00-6.

Objective. To determine that the applicant:

1. Exhibits knowledge of the elements related to thermal soaring.
2. Recognizes the indications of, and the presence of, a thermal.
3. Analyzes the thermal structure and determines the direction to turn to remain within the thermal.
4. Exhibits coordinated control and planning when entering and maneuvering to remain within the thermal.
5. Applies correct techniques to re-enter the thermal, if lift is lost.
6. Remains oriented to ground references, wind, and other aircraft.
7. Maintains proper airspeeds in and between thermals.

B. TASK: RIDGE AND SLOPE SOARING

REFERENCES: FAA-H-8083-13; AC 00-6.

Objective. To determine that the applicant:

1. Exhibits knowledge of the elements related to ridge and slope soaring.
2. Recognizes terrain features and wind conditions, which create orographic lift.
3. Enters the area of lift properly.
4. Estimates height and maintains a safe distance from the terrain.
5. Exhibits smooth, coordinated control, and planning to remain within the area of lift.

6. Uses correct technique to re-enter the area of lift, if lift is lost.
7. Remains oriented to ground references, wind, and other aircraft.
8. Uses proper procedures and techniques when crossing ridges.
9. Maintains proper airspeeds.

C. TASK: WAVE SOARING

REFERENCES: FAA-H-8083-13; AC 00-6.

Objective. To determine that the applicant:

1. Exhibits knowledge of the elements related to wave soaring.
2. Locates and enters the area of lift.
3. Exhibits smooth, coordinated control, and planning to remain within the area of lift.
4. Uses correct technique to re-enter the area of lift, if lift is lost.
5. Remains oriented to ground references, wind, and other aircraft.
6. Recognizes and avoids areas of possible extreme turbulence.
7. Maintains proper airspeeds.
8. Coordinates with ATC, as appropriate.

VII. AREA OF OPERATION: PERFORMANCE MANEUVERS

A. TASK: STRAIGHT GLIDES

REFERENCE: FAA-H-8083-13.

Objective. To determine that the applicant:

1. Exhibits knowledge of the elements related to straight glides, including the relationship of pitch attitude and airspeed.
2. Tracks toward a prominent landmark at a specified airspeed.
3. Demonstrates the effect of flaps, spoilers, or dive brakes, if equipped, in relation to pitch attitude and airspeed.
4. Exhibits smooth, coordinated control, and planning.
5. Maintains the specified heading, ±10°, and the specified airspeed, ±5 knots.

B. TASK: TURNS TO HEADINGS

REFERENCE: FAA-H-8083-13.

Objective. To determine that the applicant:

1. Exhibits knowledge of the elements related to turns to headings, including the relationship of pitch attitude, bank angle, and airspeed.
2. Enters and maintains an appropriate rate of turn with smooth, proper, and coordinated control applications.
3. Maintains the desired airspeed, ±5 knots, and rolls out on the specified heading, ±10°.

C. TASK: STEEP TURNS

REFERENCES: FAA-H-8083-13; Glider Flight Manual.

Objective. To determine that the applicant:

1. Exhibits knowledge of the elements related to steep turns, including load factor, effect on stall speed, and overbanking tendency.
2. Establishes the recommended entry airspeed.
3. Enters a 720° turn maintaining a bank angle of 45°/±5°, with smooth and coordinated control applications.
4. Maintains desired airspeed, ±5 knots.
5. Rolls out on the entry heading, ±10°.

VIII. AREA OF OPERATION: NAVIGATION

NOTE: The applicant's knowledge of this AREA OF OPERATION will be evaluated through oral testing.

A. TASK: FLIGHT PREPARATION AND PLANNING

REFERENCES: FAA-H-8083-13, FAA-H-8083-25; AIM; AC 61-134.

Objective. To determine that the applicant:

1. Exhibits knowledge of the elements related to flight preparation and planning.
2. Selects and uses current and appropriate aeronautical charts.
3. Plots a course and selects prominent en route checkpoints.
4. Constructs a flight profile to determine minimum flight altitude at go-ahead points.
5. Explains method of using lift sources and speeds effectively within and between lift sources.
6. Selects available landing area.
7. Describes coordination procedures with ATC, as appropriate.
8. For self-launch, explains the factors affecting fuel consumption, range, and engine operations.

B. TASK: NATIONAL AIRSPACE SYSTEM

REFERENCES: FAA-H-8083-13, FAA-H-8083-25; 14 CFR part 91; AC 90-48; AIM.

Objective. To determine that the applicant exhibits knowledge of the elements related to the National Airspace System by explaining:

1. Basic VFR weather minimums for all classes of airspace.
2. Airspace classes and their dimensions, pilot certification requirements, and glider equipment requirements for the following—
   a. Class A.
   b. Class B.
   c. Class C.
   d. Class D.
   e. Class E.
   f. Class G.
3. Special use airspace and other airspace areas.

IX. AREA OF OPERATION: SLOW FLIGHT AND STALLS

A. TASK: MANEUVERING AT MINIMUM CONTROL AIRSPEED

REFERENCES: FAA-H-8083-13; Glider Flight Manual.

Objective. To determine that the applicant:

1. Exhibits knowledge of the elements related to maneuvering at minimum control airspeed, including flight characteristics and controllability.
2. Establishes and maintains an airspeed at which any further increase in angle of attack or load factor would result in a stall in straight or turning flight.
3. Accomplishes coordinated flight with configuration(s) specified by the examiner.
4. Adjusts the airspeed to avoid stalls in turbulent air or as bank is increased.
5. Applies control inputs in a smooth and coordinated manner.
6. Uses proper procedures to avoid stalls when raising a lowered wing.
7. Maintains heading, ±10°, during straight flight, and the desired bank angle, ±5°, during turns.

B. TASK: STALL RECOGNITION AND RECOVERY

REFERENCES: FAA-H-8083-13; AC 61-67; Glider Flight Manual.

Objective. To determine that the applicant:

1. Exhibits knowledge of the elements related to stall recognition and recovery, including the aerodynamic factors and flight situations that may result in stalls, and the hazards of stalling during uncoordinated flight.
2. Selects an entry altitude that will allow the maneuver to be completed no lower than 1,500 feet AGL.
3. Establishes and maintains a pitch attitude that will result in a stall during both straight and turning flight with and without flaps, spoilers, or dive brakes, as appropriate.
4. Maintains a bank angle of 15°/±5°, during turns.
5. Recovers promptly at the first indication of buffeting or rapid decay of control effectiveness.
6. Uses smooth and coordinated control applications throughout the maneuver.

X. AREA OF OPERATION: EMERGENCY OPERATIONS

A. TASK: SIMULATED OFF-AIRPORT LANDING

NOTE: This landing will be performed at an established airport.

REFERENCES: FAA-H-8083-13, Glider Flight Manual.

Objective. To determine that the applicant:

1. Exhibits knowledge of the elements related to a simulated off airport landing, including selection of a suitable landing area and the procedures used to accomplish an off-airport landing.
2. Performs a simulated off-airport landing without the use of an altimeter.

B. TASK: EMERGENCY EQUIPMENT AND SURVIVAL GEAR

REFERENCES: FAA-H-8083-13, Glider Flight Manual.

Objective. To determine that the applicant exhibits knowledge of the elements related to emergency equipment and survival gear, appropriate to the glider used for the practical test, by describing:

1. Location in the glider.
2. Method of operation or use.
3. Servicing and storage.
4. Inspection, fitting, and use of parachutes.
5. Equipment and gear appropriate for operation in various climates and over various types of terrain.

XI. AREA OF OPERATION: POSTFLIGHT PROCEDURES

A. TASK: AFTER-LANDING AND SECURING

REFERENCES: FAA-H-8083-13, Glider Flight Manual.

Objective. To determine that the applicant:

1. Exhibits knowledge of the elements related to after-landing and securing procedures, including local and ATC operations, ramp safety, parking hand signals, shutdown (if appropriate), securing, and postflight inspection.
2. Selects a suitable parking area while considering wind and safety of nearby persons and property.
3. Taxies to parking area and performs engine shutdown, if applicable.
4. Services the glider, if applicable.
5. Secures the glider properly.
6. Performs a satisfactory postflight inspection.
7. Completes the prescribed checklist.

Made in United States
Troutdale, OR
11/03/2026

56844440R00031